Henry Shaw

The Rose

Historical and Descriptive

Henry Shaw

The Rose
Historical and Descriptive

ISBN/EAN: 9783337420017

Printed in Europe, USA, Canada, Australia, Japan

Cover: Foto ©berggeist007 / pixelio.de

More available books at **www.hansebooks.com**

THE ROSE.

HISTORICAL AND DESCRIPTIVE;

GATHERED

FROM VARIOUS SOURCES.

BY HENRY SHAW,

TOWER GROVE.

ST. LOUIS:
R. P. STUDLEY & Co., Printers and Lithographers.
1882.

ᵀHE ᴿOSE.

RHOS, *red, in the language of Armorica, (a province of ancient Gaul)*.
RHOOD, *red, Celtic.*
RHODON, *Greek.*
ROSA, *Latin, Italian, and Spanish.*
RUOSE, *Bohemian.*
ROSE, *English and German.*
NARD, *Arabic.*

HISTORICAL.

THE ROSE has been the favorite flower from time immemorial among the civilized nations of Europe and Asia. The cradle of the original race of man is its country, and the flowery hills which support the frowning chain of the Caucasus are adorned by this charming shrub, and gives the name Caucasian to the finest race of the human species. Many ages ago Anacreon sang the praises of the Rose.

He calls it "the most beautiful of flowers,"
"the delight of the gods," "the favorite of
the Muses," and since that time it has been
denominated the Queen of Flowers. It is frequently spoken of in Holy Writ, and Homer
often refers to the Rose, both in the Iliad and
the Odyssey. It may be said to be the oldest
of celebrated flowers, and, in the impassioned
strains of the ancients, we find it associated with
the Lily of the Valley, as expressive of all that
is pleasing to the senses, and renovating to the
mind.

The Rose, the emblem of beauty and the pride
of Flora, reigns Queen of the flowers in every
part of the globe, and the bards of all nations
and languages have sung its praises. Yet, what
poet has been able, or language sufficient to do
justice to a plant that has been denominated
the Daughter of Heaven, the glory of spring,
and the ornament of the earth? As it is the
most common of all that compose the garland
of Flora, so it is the most delightful. Every
country boasts of it, and every beholder admires
it. Poets have celebrated its charms without

exhausting its eulogium; for its allurements
increase with familiarity, and every fresh view
presents new beauties and gives additional delight. Hence it renovates the imagination of
the bard, and the very name of the flower gives
harmony to his numbers, as its odours give sweetness to the air. To paint this universal emblem
of delicate splendor in its own hues, the pencil
should be dipped in the tints of Aurora, when
arising amidst her aërial glory. Human art
can neither colour nor describe so fair a flower.
Venus herself feels a rival in the Rose, whose
beauty is composed of all that is exquisite and
graceful. Thus Roscoe tells us that

> "As Venus wandered midst the Idalian bower,
>   And watched the Loves and Graces round her play,
>   She plucked a musk rose from its dew-bent spray,
> And this, she cried, shall be my favorite flower;
> For o'er its crimson leaflets I will shower
>   Dissolving sweets to steal the soul away."

It has been made the symbol of sentiments
as opposite as various. Piety seized it to decorate
the temples, while Love expressed its tender-
ness by wreaths; and Jollity revelled adorned
with crowns of roses. Grief strews it on the
tomb, and Luxury spreads it on the couch. It

is mingled with our tears, and spread in our
gayest walks; in epitaphs it expresses youthful
modesty and chastity, while in the song of the
Bacchanalians their god is compared to this
flower. The beauty of the morning is allegori-
cally represented by it, and Aurora is depictured
strewing roses before the chariot of Phœbus:

> "When morning paints the orient skies,
> Her fingers burn with roseate dyes."

The Rose is supposed to have given name to
the Holy Land, where Solomon sang its praises,
as Syria appears to be derived from Suri, a
beautiful and delicate species of Rose, for which
that country has always been famous; and hence
Suristan, or the "Land of Roses." The island of
Rhodes owes its name to the prodigious quantity
of roses which formerly grew upon its soil.

Anacreon's Birth of the Rose stands thus
translated by Moore:

> "Oh! where could such a plant have sprung?
> Attend—for thus the tale is sung:
> When hurried from the watery stream,
> Venus appeared in flushing hues,
> Mellowed by ocean's briny dews—
> When in the starry courts above,
> The pregnant brain of mighty Jove

Disclosed the nymph of azure glance,
The nymph who shakes the martial lance!
Then, then, in strange eventful hour,
The earth produced an infant flower,
Which sprang, with blushing tinctures drest,
And wantoned o'er its parent's breast.
The gods beheld this brilliant birth,
And hailed the rose—the boon of earth!
With nectar drops a ruby tide,
The sweetly-orient buds they dyed,
And bade them bloom, the flowers divine
Of him who sheds the teeming vine;
And bade them on the spangled thorn
Expand their bosoms to the morn."

The first Rose ever seen was said to have been
given by the god of love to Harpocrates, the god
of silence, to engage him not to divulge the
amours of his mother Venus; and from hence
the ancients made it a symbol of silence, and
it became a custom to place a Rose above their
heads, in their banqueting rooms, in order to
banish restraint, as nothing there said would
be repeated elsewhere; and from this practice
originated the saying "sub rosa" (under the
rose) when anything was to be kept secret.

In the mythologic ages it was sacred as the
flower of young affection and endearment, and
of mature love—the flower of Cupid and Venus,

and stripping this of the mythological phrase-
ology, which in all cases was a fictitious mantle
thrown around something previously felt, no
similitude of any flower could be more ap-
propriate. The rosebud, the sweetest object that
appears in the garden, is typical of all begin-
nings from the issue of which joy and pleasure
may be expected.

> " Ah, see the Virgin Rose, how sweetly shee
> Dost first peep forth with bashful modestie,
> That fairer seems, the less you see her may!
> Lo! see soon after, how more bold and free
> Her bared bosom she doth broad display!
> Lo! see soon after how she fades and falls away!"
>
> *Spenser's Faerie Queen*—1589.

The early dawn; young schemes and projects;
young life; young love, and a hundred other
associations, all of a delightful kind, are
associated with the Rosebud. There seems a
physical attraction in it beyond all flowers in
every stage of its growth, and an attraction
which addresses itself strongly to the feelings.
When roses are in full bloom, they certainly
are the most delightful flowers the amateurs can
cultivate; the Rose is the most obedient to his
labor, and rewards the cultivator richly for his

care and skill. Still, there are persons who share with the black beetle a positive dislike for the Rose. Among those who have taken so prominent a part in public life as to have attracted the attention of history, is the famous Chevalier de Guise, who could not smell a rose without feeling uncomfortable; and Vinieri, one of the Doges of Venice, suffered under the same inconvenience for the enjoyment of the garden. Anne of Austria, wife of Louis XIII, could not even look at a rose in a picture without being seized with tantrums.

In the East there is still the belief that the first Rose was formed by a tear of the Prophet Mahomet, but nations of more cool and dispassionate imaginations have sometimes admitted that its origin was lost in obscurity. Roses were used in very early history among the most potent ingredients of love philters. They seem to have been imported by the Romans from Egypt until the age of Domitian. Antiochus slept upon a bed of rose leaves; Mark Anthony begged that Cleopatra would cover his tomb with these flowers, and *mea rosa* was a favorite

term with Roman lovers. Homer has adorned
the shield of Achilles, and the helmet of Hector
with roses. Among the Greeks it was customary
to leave bequests for the maintenance of rose
gardens over the grave of the testator, and at
Torcallo, near Venice, an inscription may still
be seen, which shows that the fashion was
adopted in Italy. In Turkey, a stone rose is
often sculptured above the graves of unmarried
women. A charming bas-relief on the tomb of
Madame de la Live, who died at the age of
twenty-one, represents Time mowing a Rose with
his scythe.

According to Indian mythology, Pagodastri,
one of the wives of Vishnu, was found in a rose.
Zoroaster is said to have made a rose-tree spring
out of the earth and bud and bloom in the
presence of Darius, who had called upon him
to perform a miracle. In one of the books at-
tributed to Solomon, eternal wisdom is compared
to rose-trees at Jericho. Princess Noumahal, the
most lovely lady in the harem of the Great
Mogul, had a canal filled with rose-water, and
rowed about in it with her august consort, the

heat of the sun disengaged the essential oil from
the water, and their majesties having observed
the fact, invented Otto of Roses. When the
Soldan, Saladin, who had so much trouble with
our hard-fisted King Richard, and his turbulent
Christian friends, took Jerusalem in 1188, he
would not enter the temple, which he profanely
called a mosque, until he had its walls washed
with rose-water; and Samet assures us that five
hundred camels were no more than sufficient to
carry the purifying fluid. Also after the taking
of Constantinople by Mahomet II, 1455, the
church of Saint Sophia was solemnly purified
with rose-water before it was converted into a
mosque. The high priest of the Hebrews wore
a crown of roses when he offered up certain
sacrifices under the Mosaic dispensation, and it
was perhaps in remembrance of this fact, that
the Synod of Nismes, which was held in the
third century, enjoined every Jew to wear a
rose on his breast, as a distinguishing mark of
his inferiority. In many countries, the Jews still
celebrate the festival of Eastern Flowers; during
which they ornament their lamps, chandeliers
and beds with roses.

When Maria Antoinette passed through Nancy,
on her way to be married to Lewis XVI, the
ladies of Lorraine prepared her a bed strewed
with roses. In the middle ages roses were held
so precious in France that a royal license was
necessary to grow them; Charlemagne, recom-
mended the cultivation of the Rose in his
"Capitulation." The Persians of Shiraz stop
their wine bottles with roses to give the wine
a pleasing perfume; and during the festival of
Abrizan, which takes place during the equinox,
Persian ladies throw roses at each other when
they visit. "On entering the gardens of the royal
palace of Persia," says Sir Robert Porter, "you
are struck with the appearance of rose-trees full
fourteen feet high, laden with thousands of roses,
blooming and diffusing a delicacy of perfume,
that imbued the whole atmosphere; but in these
delicious gardens of Negaristan, the eye and the
olfactories are not the only senses regaled by
the presence of the Rose; the ear is enchanted
by the wild and beautiful notes of multitudes
of nightingales, whose warbling seems to increase
in melody and softness with the unfolding of
their favorite flowers. Here, indeed, is the

genuine country of the nightingale and the Rose.

At Rome the Golden Rose was consecrated by the Pope and given to some prince or princess as a mark of the Sovereign Pontiff's favor. Urban V gave the Golden Rose to Joan, Queen of Sicily, in 1368. Henry VIII of England received a Golden Rose from Julius II, and from Leo X. Roses were often, in the days of chivalry, worn by the cavaliers in tournaments, as an emblem of their devotion to love and beauty.

DESCRIPTIVE.

THE habits and colours of the several varieties of the Rose are almost without end, and yet there is great beauty in each of them. Then the perfume with which they embalm the zephyr, as it plays gently over them, diffusing an odour most delightful to the sense, and exhilarating to the mind. In most instances the odour of a flower dies along with it, but not so with the Rose, for some leaves gathered by the writer at Tower Grove, in 1852, and preserved in a jar, are now (1877) still fragrant. We find it yielding a variety of fragrant liquors, and the attar of roses especially, when prepared in the valley of the Ganges, or in Cashmere, where square miles are devoted to the growth of this flower, is now almost the only substance which, weight for weight, is more valuable than gold.

The shrub varies in size, usually from one to six or eight feet; the colours are red, white, yellow, purple-striped, and in almost numberless shades

and varieties; the flowers single, semi-double
and double; the odor is universally grateful; the
green rose is a monstrosity, without fragrance.

The Rose is cultivated in every garden, from
that of the most humble peasant to that of persons of rank and wealth, but will not grow to
perfection in the smoky, dusty atmosphere of
large towns. Some species, as *Rosa centifolia,
damascena, etc.*, are also cultivated by commercial gardeners, on a large scale, for distilling
rose-water, or for making attar or essential oil of
Roses; six pounds of rose-leaves will impregnate,
by distillation, a gallon of water strongly with
their odour, but a hundred pounds affords scarcely
half an ounce of attar. This most delicious of all
perfumed essences is obtained by the simple
distillation of rose-leaves. In our climate Roses
are not sufficiently scented to produce the odor-
iferous essential oil. Among the most favorable
countries for the production of the most highly
scented Roses is the middle portion of European
Turkey, at the base of the southern slope of the
Balkan mountains, in localities where the Roses
are protected against all winds except those from

the south; and the flowers thus attain a luxuriance of perfume and growth peculiar to these
favored regions. The center of the cultivation
and distillation of the Rose is the town of
Kazanlik, situated in the province of that name,
and is watered by many mountain streams that
furnish a suitable water for the distillation of the
precious attar. The numerous villages of the
province, inhabited by Turks and Christians
employed in the cultivation of the Rose, all live
in peace together and prosper; finding by ex-
perience that it is better and more wise to work,
than to waste time in religious and political
quarrels. The great harvest commences May
15th, and lasts until June 5th or 10th; the gather-
ing is done in the morning before sun-rise, so as
to have the benefit of all the fresh perfume of
the flowers, which might be drawn off by the
heat of day. Every Rose farmer has his own
small, roughly constructed stills for producing
the otto or attar immediately after picking the
flowers; and thousands of industrious workers
are thus occupied, earning in the single short
period of twenty days the product of a year's
labor, cultivating and taking care of the grow-

ing plants. The total yearly production of the
province of Kazanlik is from 3,500 to 6,000 pounds,
the product of 1866, but in 1872 only 1,700 pounds
could be obtained. When distillation is over, the
farmers come to the great commercial centres of
Constantinople and Adrianople to sell their
products.

Unfortunate Kazanlik! ravaged by the horrors
of war; in place of quiet villages reposing in the
valleys of the Balkans, presents at this time
(1877) a scene of ruin and devastation; the dwell-
ings of the inhabitants, the churches of the
Christians, and the mosques of the Moslem, are
now heaps of smouldering ruins. The Rose
cultivators slaughtered, fled, or suffering from
pestilence and famine; the Rose gardens, once so
delightful, are now overrun by hordes of Musco-
vite soldiers, or serve as pens for the horses of
the Cossack.

THE GEOGRAPHY OF THE ROSE.

ROSES in a wild state are natives of Persia, India, China, Barbary, Europe, and North America, and confined to the northern hemisphere, never having been found wild very near to, or south of the Equator. The vast continent of Australia, rich in botanical treasures as it is, has not revealed to us a single species.

Certain authors assign the provinces of Georgia and Circassia as the native places of the older roses; various countries possess species and varieties that are peculiar to them. The *Rosa Polliniana* is peculiar to Mount Baldo in Italy, *R. Lyonii* to Tennessee, while the *R. arvensis* or Field Rose is to be found in all the countries of Europe, and the *R. canina* or Dog Rose in Europe, as well as in a considerable portion of Asia.

AMERICA.

The Roses of this continent are *R. blanda* (the early wild Rose), found among the rocks and

banks of the Northern States, and on the glaciers
of the northerly regions of the New World, its
bright corolla unfolding itself immediately on
the melting of the snows. This shrub is found
as far north as the frozen deserts, between 70°
and 75°. *R. Hudsonensis* is found on the shores
of Hudson's Bay, within the polar circle, where
it produces clusters of double pale flowers. *R.
fraxinifolia* has small red, heart-shaped petals,
and is found in Newfoundland and Labrador. *R.
nitida* (wild shining dwarf rose), which has deep
red flowers, abounds on the Northern coasts, and
is used by the Esquimaux for decorating their
hair and seal skin dresses. *R. lucida* (Ehrhart,
low wild Rose) is found in the marshes of Carolina, in Missouri and westward to Nebraska. *R.
Woodsii* on the Banks of the Missouri and *R.
Carolina* (Linnæus), Swamp rose in the adjoining
marshes. *Rosa setigera* (the Prairie Rose), a fine
species, the only American climbing rose; several
varieties and forms in cultivation. The rose of
Mexico is *R. Montezuma*, a sweet scented thorn-
less species, which abounds in the highest parts
of Cerro Ventoso near San Pedro, where it was
discovered by the learned travellers Humboldt

and Bonpland in lat. 19° North. The total number of American roses are estimated at fourteen.
Those of France are twenty-four, and of Britain
near the same number.

ASIA

Has a greater variety of species of the Rose than
the rest of the earth united; thirty-nine, that
admit of accurate definition, having been already
established. Of these, the vast empire of China,
where both agriculture and horticulture are arts
in high estimation, has a claim to fifteen. First
the *Rosa semperflorens* (the Ever-blooming), flowers scentless, of a pale pink color, producing
a pleasing effect when half grown. *R. Lawren-
ciana* (the Fairy Rose), a beautiful little shrub,
but unlike most other dwarfs of the vegetable or
animal kingdom, perfect in symmetry and pro-
portion. *Rosa multiflora* attains a growth of 12 to
15 feet, forming beautiful bouquets on the tree.
Rosa Banksiana extends its flexile branches over
rocks and hillocks, bearing a profusion of small
very double yellow flowers, remarkable for their
violet-scented fragrance. *Rosa microphylla* (small
leaved Rose) is a favorite garden shrub of the

Chinese under the name of Haitong Hong, having small double pink flowers, and foliage of peculiar delicacy, and *Rosa bracteata*, the Macartney Rose, is a distinct species. Japan, between the 30th and 40th degrees of latitude, has all the roses of China, besides a peculiar species (hardy at St. Louis). The *Rosa rugosa*, with crimson flowers and large handsome bright fruit.

India has the *R. Lyellii*, remarkable for its profusion of milk-white flowers. *Rosa Brunoni*, with snowy-white petals, ranks high among the roses of India. The parched shores of Bengal are covered during spring with a beautiful white rose; also found in China and Nepaul; while in the vast thickets of the beautiful *Rosa semperflorens* the tigers of Bengal and the crocodiles of the Ganges are known to lie in wait for their prey. In the gardens of Kandahar, Samarcand and Ispahan the *R. arborea* (the Tree Rose) is cultivated extensively by the Persians. The *R. Damascena*, transported from Damascus by the Crusaders, affords our gardens an infinite number of beautiful varieties, and adorns the sandy deserts of Syria with its sweet and bright tinted flowers.

Towards Constantinople the *R. sulphurea* displays its double flowers of a brilliant yellow. The North-West of Asia, which has been signalized as the Father-land of the Rose-tree, introduces to our admiration the *Rosa centifolia*, the *Hundred-petal rose*, the most esteemed of all, and celebrated by the poets of every age and country, and with which the fair Georgians and Circassians adorn their persons. The *R. ferox* mingles its large red blossoms and thorny branches with those of the centifolia or hundred-leaved, and the *R. pulvurulenta* is also seen on the peak of Nargana, of the Circassian chain. In the north of Asia, Siberia boasts of three species; advancing further north in the Russian provinces ten or twelve other species are found, in particular the *R. Kamschatica*, bearing solitary pink flowers.

AFRICA.

On the plains towards Tunis is found *Rosa Moschata*, the Musk rose, whose tufts of white flowers give out such a musky exhalation. This charming species is also found in Egypt, Morocco and the Island of Madeira; in Egypt also grows the *Rosa canina* or *Dog-rose*, so com-

mon throughout Europe. In Abyssinia we find
an evergreen Rose-tree, with pink blossons, which
bears the name of the country, *R. Abyssinica.*

EUROPE.

Commencing with Iceland, a country so infertile
in vegetation, we find *R. rubiginosa* (sweet briar)
with pale solitary cup-shaped flowers. In Lap-
land (honored by the researches of Linnæus
himself), blooming almost under the snows of
that severe climate, grows the *R. majalis,* small,
sweet, and of a brilliant color; and the same
beautiful species enlivens the cheerless, rude climate of Norway, Denmark and Sweden The
sweet briar, the May-rose, the *Cinnamon rose,* the
small red flowers of which are sometimes double, as well as several other hardy species, may
be found in the countries of Northern Europe.

Six species are indigenous to England. *R.
involuta* exhibits its dark foliage and large red
flowers amid the forests of North Britain; in the
same districts are found *R. Sabini, the R. villosa*
with crimson flowers, and *R. canina,* the common
Dog-rose.

The environs of Belfast produce the *R. Hibernica*, thought to be a variety of *R. canina*, growing in loamy land.

Germany, though unproductive in Rose-trees, has several highly curious species in *R. turbinata*, and *R. arvensis* with large flowers, red and double in a cultivated state.

The Swiss mountains, and the Alpine chain in general, are rich in native roses. Besides the *Field rose, R. arvensis*, they have *R. Alpina*, an elegant shrub, with red solitary flowers, furnishing many varieties in cultivation. Among the Alpine roses the *R. rubrifolia*, of which the red tinted stems and leaves, as well as the pretty little blossoms of a deep crimson tint, form an agreable variety to the verdure of the surrounding foliage.

In Greece and Italy the *R. glutinosa*, of which the leaflets produce a viscous matter, the flowers being small, solitary, and of a pale red. Italy has several species, among which are *R. Polliniana*, with fine, large, purple flowers growing in clusters, and found in the neighborhood of Verona.

The *R. moschata* and *R. Hispanica* flourish in
Spain. The *R. sempervirens*, common in the Bale-
aric Islands, grows spontaneously throughout the
South of Europe and Barbary ; its foliage of a
glossy green is intermingled with a profusion of
small white, highly scented flowers.

In France nineteen species are claimed by the
Flora of Decandolle; in the Southern provinces
R. eglantina, the *Yellow Briar*, is found, whose
golden petals are sometimes varied into a rich
orange. In the forests of Auvergne we find the
R. cinnamomea, which derives its name from the
colour of its branches, the flowers being small and
red. The *R. parviflora* or *Champagne rose*, a
beautiful miniature shrub, adorns the fertile valleys in the vicinity of Dijon. The *R. Gallica* is
one which has afforded varieties of every hue,
more especially the kinds known as the Provins
roses, white, pink and crimson. In the Eastern
Pyrenees grows *R. moschata*, a beautiful variety
of which is known as the *Nutmeg Rose* and *R.
alba* in the hedges and thickets of various De-
partments.

CULTIVATED SPECIES AND VARIETIES
OF THE ROSE.

HUNDRED-LEAVED.—*Provins or Cabbage Rose.—Rosa centifolia.*—Its name Provins from a town twenty leagues from Paris; the French also call vines grown from layers "Provins," and Cabbage, the English name, from the form of the rose; blooming annually; a well known and popular species, cultivated in English gardens for more than 300 years.

This is the Rose with which painters choose to represent Love and Hymen, and is selected as the emblem of Grace.

"EMILIA—   Of all flowers,
        Methinks a Rose is best.

SERVANT—   Why, gentle madam?

EMILIA—It is the very emblem of a maid :
        For when the West wind courts her gently,
        How modestly she blows. and paints the sun
        With her chaste blushes! When the North comes near her,
        Rude and impatient. then like Chastity.
        She locks her beauties in her bud again,
        And leaves him to base briars."
                                BEAUMONT.

Bieberstein asserts to having seen it growing on the Caucasus; and is the kind supposed to have been mentioned by Pliny, as a great favorite among the Romans. It can vie with all others in size, beauty, perfection of form and fragrance. In the humid air of Britain it blooms for two months in summer, and is found around almost every cottage; but with us a week in June displays every flower, and in hot weather they flower and bloom in a day or two. The common Cabbage Rose is of a clear delicate pink color, and of several varieties. The *Unique* or *White provins* is of a pure white.

Rosa muscosa, the *Moss Rose*, is also a variety of the Provins rose, which when grown in Italy loses its mossiness; color of various tints, of red and crimson. The *white moss* is much admired, and also the *Bourbon moss*, growing 6 to 8 feet in height. The Moss rose is made the emblem of voluptuous love, and the creative imagination of a German poet thus pleasingly accounts for this Rose having clad itself in a mossy garment:

> " The angel of the flowers one day
> Beneath a rose-tree sleeping lay,
> That spirit in whose charge is given
> To bathe young buds in dews from heaven.

Awaking from his light repose,
The angel whispered to the rose,—
'Oh, fondest object of my care,
Still fairest found where all are fair,
For the sweet shade thou'st given to me,
Ask what thou wilt, 'tis granted thee.'
Then said the rose, with deepened glow,
'On me another grace bestow.'
The spirit paused in silent thought—
What grace was there that flower had not?
Twas but a moment— o'er the rose
A veil of moss the angel throws;
And, robed in nature's simplest weed,
Can there a flower that rose exceed?''

Rosa Gallica, the *French Rose*, the distinguishing features of this family are its strong, upright flowering stalks: the colors vary from pink to the deepest shades of crimson. It also includes the the striped, mottled, and variegated roses; the varieties of *R. Gallica* are very numerous.

Rosa alba, the *White Garden Rose*, often called the White Climbing rose, may be considered a hardy Rose.

Flora's Lexicon says of the White Rose-bud: Before the breath of love animated the world, all the roses were white, and every heart insensible. An old English poet (Herrick) says,

"As Cupid danced among
The Gods, he down the nectar flung:
Which on the white rose being shed,
Made it for ever after red.''

Another poet makes the Rose to say :

> " 'Twas from Love I borrowed, too,
> My sweet perfume, my purple hue."

And by another poet we are indebted to our first mother for colored Roses :

> " As erst, in Eden's blissful bowers,
> Young Eve surveyed her countless flowers,
> An opening rose of purest white
> She marked with eye that beamed delight.
> Its leaves she kissed, and straight it drew
> From beauty's lip the vermil hue."

Rosa Damascena, *Damask Rose*, transported from Damascus to Europe by the Crusaders, has that delicious odour so peculiar to the species, having a succession of blooms. The monthly Damask is a bright pink, blooming in clusters, and repeatedly during the season, if in rich soil, and is a general favorite. Painted Damask has the quality indicated by its name, but not so well painted as the old York and Lancaster (*R. Gallica*) which is often striped, and frequently one half pink and the other half white, thus according with the tradition, that, on the extinction of the feud between the houses of York and Lancaster, this rose sprung up, with one side pink, and the other white.

Rosa spinosissima. The *Scotch* or *Burnet* Rose takes its name from being very thorny and has been found growing in Scotland, and in many of the Alpine districts of Europe; and known as the Scotch rose from the fact of the first introduction of it in a double state having been by Mr. Robt. Brown, Nurseryman of Perth, Scotland, about the year 1793, who introduced it into his nursery by seeds from a neighboring hill; and by raising new plants every year, in 1803, had good double specimens, which have since been diffused over the world. This patriarch of horticulture afterwards became domiciled in Philadelphia, and to him the late Mr. Robt. Buist, Sen., owed (as he says) many practical facts, where he died at an advanced age, in 1845, and lies interred in Philadelphia cemetery. The original varieties of *R. spinosissima* are not much cultivated in this country, but are classed under the head of Perpetual Roses.

CLIMBING ROSES.

Prairie Rose, R. setigera, a native rose remarkable for its perfectly hardy growth, and will bear without injury the icy brezes of the St. Law-

rence and the melting vapors of the Mississippi.
There are several varieties, as *Queen* of the *Prairies*, *Superba*, &c., the flowers are produced in large
clusters of various shades of color from blush to
deep rose. *Baltimore Belle,* nearly white and of
perfect form; but all are devoid of fragrance
or nearly so. *Boursault Rose, R. Alpina,* takes
its name from Mons. Boursault, a distinguished
French horticulturist. Flowers of a reddish purple color, of several varieties, some of which are
only semidouble.

Ayrshire, R. subdecidua, Tea scented flowers,
compact and perfectly double, of several varieties, and hardy, variety of *R. sempervirens*
Rosa multiflora, pink, and *R. Banksia* with yellow flowers, are natives of China, and tender
at St. Louis.

HYBRID CHINA ROSES.

From seeds of Bengals, Teas, &c., impregnated
with pollen from Centifolia, Damask, and other
sorts, that bloom only once in the season and
hardy at St. Louis; some are of luxuriant growth,

others are dwarf, and compact in habit. In this class are ranked

Brennus—very bright red.

Cerisette—very pretty, flowering profusely.

Coupe d'Hebè.—Hebe's Cup, large and perfect form.

George IV.—a splendid variety raised by the late Mr. Rivers of England, from seed, a great many years ago.

Madame Plantier—pure white in clusters.

Velour episcopal—velvety crimson.

Louis Philippe—a splendid dark rose, blooms in great abundance, almost perpetually, with the fragrance of the Damask. The Hybrid Chinas bloom once a season and are sometimes wrongly classed with the Hybrid Perpetuals, or Remontantes.

BRIAR ROSES.

The *Sweet Briar, Rosa rubiginosa.* The eglantine of the poets, well known for the delicious sweetness of its foliage and flowers.

> "The sporting sylphs that course the air,
>   Unseen on wings that twilight weaves,
>   Around the opening rose repair,
>   And breathe sweet incense o'er its leaves."

Rosa lutea, the *Yellow Austrian Briar,* is very common in old gardens.

Persian Yellow Rose, of a brilliant golden color, cup shape and quite double.

Harrison's Yellow Sweet Briar. This pretty yellow Rose was grown by a Mr. Harrison, near New York, some forty years ago, blooming early and in great profusion.

Roses that bloom the whole season are

Rosa Indica, Bengal or *Daily Rose,* is of many varieties, of rather a hardy nature, but will not stand the winters of St. Louis without protection; it can be propagated by cuttings, and grows rapidly in the warm months of July and August. The varieties of a pale tint to dark crimson are too numerous to record. Agrippina, crimson, and Mdme. Bosanquet, creamy-blush, are old and favorite Roses.

Rosa Indica odorata, Tea scented Rose, was introduced from China to Europe some 70 years since. The varieties from it are magnificent; many of them of the most luxurious character, with flowers of all shades from white to red, requiring the protection of a frame or green-house in winter; planted out in fine weather in April, every lover of these celebrated roses can enjoy them in the greatest luxuriance from June to October. Of this class are the popular Marshal Neil, Duchess of Edinburgh, and many others.

Noisette Rose, this famous Rose originated in Charleston, South Carolina, with Mr. Noisette in 1815, who sent it to his brother, a then celebrated nurseryman of France, and created a very great excitement; from it thousands of seedlings have been raised. It flowers from June to October, in immense clusters of large flowers, requiring a slight protection of litter or leaves in winter, *Aimée Vibert, Lamarque,* and other celebrated roses, are of this class.

Isle of Bourbon Rose, R. Bourboniana. Some of the most beautiful of autumnal roses belong to this class, the Autumn indeed being their peculiar season. The only roses known on the Island of Bourbon were the common China, and the Four Seasons, till about 1816, when a Monsieur Perichon was planting a hedge of these. Among the plants he found one very different from the others in its shoots and leaves, which induced him to plant it in his garden, where it was discovered by a French botanist, and sent home in 1822 to Mons. Jacques, then gardener at the Chateau de Neuilly: and having thus attracted the attention of the leading rose growers of Paris, they set to work and propagated it extensively. From July to September they are constantly in bloom at Tower Grove. Among the favorite Bourbons are Hermosa, Queen of the Bourbons, and Souvenir de la Malmaison.

Hybrid perpetual or *Remontante Roses*, a leading family of hardy roses of every variety of color from pure white to velvet crimson. This tribe originated some forty years ago with Mr. Laffay on the hill of Modon, near Paris, a little spot of ground where he produced the very finest of the family. Another race of hybrids between the remontantes and the Bourbon has produced results of superior quality; their general habit is robust and vigorous, their flowers large, fragrant, and almost of every color, such as Anna de Diesbach, General Jacqueminot, Giant of Battles, and others among the finest in our garden.

Rosa Laurenciana, *Fairy Rose*. These diminutive roses were first introduced from China, where the greatest efforts of Horticulture are directed towards dwarfing every tree, or shrub. These beautiful little plants of sixteen varieties are well worthy our attention, from their dwarfness and perfect symmetry of form, often flowering when not more than six inches high, and for the beautiful color of their diminutive rosebuds. They are named in honor of Miss Lawrence, who published in London, 1810, a collection of engravings of the Rose, accurately drawn and elegantly colored.

Musk scented Rose, *Rosa moschata*, named from the peculiar and agreeable odor it exhales in the evenings of the cool autumnal months, which is the season it flowers most abundantly—

"When each inconstant breeze that blows,
Steals essence from the musky rose."

It was formerly much valued for its fragrance, when musk was a fashionable perfume.

Small leaved Rose, *R. microphylla*, a cultivated variety from China; was introduced from Canton into the Calcutta Botanic gardens by Roxburgh, and from thence diffused generally into Indian gardens; of luxurious growth, with small leaves of a lively green, flowers double and semidouble, rose color and light pink; supposed to have originated from the old Macartney rose; stands the winters of St. Louis, with a protection of dry leaves.

Macartney Rose, Rosa bracteata, with bright, thick evergreen leaves, a white rose, native of China.

Evergreen Rose, R. sempervirens, a white climbing rose, of the south of Europe, not cultivated in this latitude.

Loudon, in Encyclopédie of Plants, describes 97 distinct species, and gives the names of 468 varieties that he had seen in cultivation in 1840.

CULTURE OF ROSES.
BRIEF ITEMS.

Situation. A place apart from other flowers should be assigned to them, if possible, sheltered from high winds, but open and not surrounded by trees, as closeness is very apt to produce mildew; also dust and smoke should be avoided.

Soil. A most important item in their successful cultivation. That they especially delight in a rich loamy soil; when heavy, good drainage is most important, with a small addition of coal ashes.

Propagating. Budding, layering, and cuttings, are the usual modes; plants on their own roots are generally preferred. Standards are with difficulty kept alive in the hot summer season.

Planting. When the seasons are favorable October and April are preferred; and before cold weather sets in, give winter protection with dry leaves and a little soil.

Manuring. Roses are strong feeders, and will take almost any amount of manure; cow dung or well rotted stable manure are generally preferred. A good top-dressing may be laid on the beds in autumn, and in spring dug-in.

Watering. When coming into bloom, if the weather is dry, give a good drenching twice or three times a week. In May and June syringe with sulphur and lime to prevent mildew and damage by caterpillars.

Pruning. May be done after the beginning of March, according to the season; cut out all wood

over two years old, and all weakly shoots, and all
delicately growing kinds should be cut down to
3 or 4 eyes. Stronger growing kinds may be left
longer. Teas and Noisettes require less cutting
back, the tips should be shortened, and weak
shoots cut out, and they should not be pruned till
May. Use a good pruning knife in preference to
a secateur; it cuts cleaner, and does not bruise
the wood.

The real cause of the eminence of France in the
cultivation of Roses is the fact that it absorbs the
almost exclusive attention of French florists; the
high price of fuel in France places the cultivation
of tender exotics almost out of the question.

The first impulse was given to the culture of
the Rose in France at the commencement of the
present century, under the auspices of the Em-
press Josephine, who caused her own name to be
traced, at considerable expense, in the parterres
of Malmaison, with a plantation of the rarest
roses. The Rose School of the Luxembourg nur-

sery is second only in national importance to the School of Vines.

Remarkable as Rose growers, Noisette has given his name to a beautiful and prolific variety; obtained in the first instance at Charleston, South Carolina, by his brother Philip Noisette. Having amassed a considerable fortune, the Noisettes no longer after 1835 continued to raise roses from seed; and that branch of cultivation was carried on at Paris by Laffay, a most enthusiastic and intelligent gardener; and by Vibert, who has written some valuable treatises on the culture of roses. Cels and Sisley-Vandael exported largely; the latter excelled in the production of the Tea or scented China rose. Boursault's celebrated collection fell to decay; while that of Decemet of St. Denis, one of the first growers, who attained much distinction, was cut up by the invading troops in 1814. The same branch of rose culture was practised with great success at Brussels and Dusseldorf. In the imperial gardens at Monza near Milan 39 varieties of the China rose have been obtained by the celebrated Nillarese, and Genoa, Marseilles and Avignon have added to the number.

The China or Bengal Rose sent to the Botanic garden at Kew, 1780, from Canton by a botanist named Ker, did not reach France till the year 1800.

In addition to the interest excited by his seed-lings, the attention of the rose-growers is eagerly directed to the accidental varieties produced by what is called a "sport," or a branch losing the habit of the plant on which it grows, and assuming new specific characters. In this way the Moss unique was originated at Clifton, and the beautiful Rose *cristata* in Switzerland; and more recently the charming Tea-rose Isabella sprunt in North Carolina. The Ayrshire roses were chiefly obtained from seed at Dundee in Scotland, and the yellow sweet-briar at Pitmaston. To enter into the origin of even the finer modern varieties would however be an endless task. The most scientific work which has appeared in England on Roses is the Rosæ Monographia of Prof. Lindley, 1819, in which above a hundred species and sub-species are described.

In France, Redonté and Thory published a splendid work in folio, entitled "Les Roses," containing plates of the species and varieties of this flower, and a "Monographie du genre Rosa" in 1820. The Rose Manual by the late Mr. R. Buist, Philadelphia, is one of the best of American works on the Rose, although many new and fine varieties have been introduced since Mr. Buist, Sr., wrote.

A History of Roses by a French Rosarian, 1874.

SIXTY years ago the Rose list of Decemet was considered very full; it included 300 roses. In 1830 we knew about 2,500 varieties; we have now more than double that number, and this fact is mainly owing to the seedlings by hybridisations, and the intelligent selections of French florists.

If France is not the native country of the Rose, which like the vine was born in Asia, it is in France that the vine and the Rose have found the soil, the climate and the care which have made their fortunes. More than 5,000 Roses! and how many simple admirers of nature, with the poets at their head, seem to think them only one; the Rose of Homer, of Virgil, of Delille, and of St. Lambert. But of the poet's Rose we have no picture, no actual record. There is every reason to suppose however, it was the *Rosa centifolia*, the

Hundred-leaved Rose, which the poets sang, and it was almost the only one to which the painters paid homage; since Redouté (a celebrated painter of flowers) their pencils have delineated many of the lovely varieties which our gardens have produced; but look at the paintings of the 16th, 17th and 18th centuries, and you will see none but the Cabbage (100-leaved) Rose, the English White and Red Roses of York and Lancaster, and the Yellow Rose, which only became really double about a hundred years since. We must admit the Cabbage Rose, as the Dutch formed it, has never yet been surpassed by any of the productions of our florists; none had even approached it; and it has the advantage of flowering twice a year like the Roses of Pæstum, which were probably a kind of four seasons' Rose. In almost every country these roses are apparently as old as the hills; but it is from the Asiatic hundred-petaled rose tribe that man has everywhere his first delight.

The Rose of Provence, down to the time of the crusades at least, the only famous rose in France, was the first of these oriental visitors acclimated amongst us; the Rose of Damascus, which has much of their blood in its veins, was brought to France it is said by the Count De Brie, and the neighborhood of Brie, Conté Robert, is still the great field from which France supplies Europe and America with rose-trees. The old English Rose was a daughter or cousin to the Provence Rose. Their Portland itself is a species of Cabbage Rose, *Centfeuilles*—what rendered it famous was its flowering twice. Where did the Dutch find the true hundred-petaled Rose? Perhaps, like us, they got it from the Moors in Spain, or the merchants of Smyrna. Wherever they obtained the original, it was their art which developed all its beauties.

Till nearly the end of the reign of Louis the XIV the gardens of Europe depended upon the same source—improving the known varieties by grafting, without raising seedlings, and making scarcely any new acquisitions. In 1735 the Fairy Rose, Pompon, was discovered in a wood near Dijon; it was not of much beauty then. The Moss Rose, issue of the Cabbage Rose, appeared about the same time. It is thought that Miller,

the learned English gardener, obtained it in 1727.
The Countess de Genlis introduced it into Paris
about twenty years after that date.

But an unexpected era was now approaching.
All was changed when the Tea Rose reached us
from China, and the Bengal Rose from India.
These precious shrubs—near relations, however,
of the Dog Rose of our own woods—were the
richest presents that the soil of India could give
us. We possessed the finest of roses, but they
only blossomed during a few days at the end of
the spring; the new comers decorated our gardens to the end of autumn with an abundance
and freshness of foliage and flowers hitherto
unapproached.

These were, however, only half the treasures
scattered by Flora over the gardens which she
loves. The marriage of the old with the fruitful
young rose was soon consummated, and from that
time the wand of the fairy multiplied the beauties in the hands of our ablest florists. Hybrid-
isation and seedlings aiding each other, there
is scarcely a limit to the caprice of the most
daring cultivator.

It was about 1789 that the Bengal and Tea
Roses became well known. The Banksian climb-
ing Rose was only brought from China in 1807.
The Bourbon Rose appeared somewhat later; the
Noisette had then already arrived from America.

Let us not be ungrateful to our old roses; at the
very moment when an unknown field was opened
up to us in the East, the Portland, cultivated by
Mons. Telieur, of Ville-sur-Ars, or perhaps by
Mons. Souchet, gave us that admirable Rose du
Roi, so vigorous, so hardy, so well formed, so
delicious in color and perfume.

The free-flowering Rose of China bloomed for
the first time in France in 1812; the English
knew it before that date; how many names and
dates should we have to inscribe, to perpetuate
the memory of the conquests of the Rosary
during the last fifty or sixty years!

Many exquisite beauties have been brought by
the art of man from beneath the veil which
nature had thrown over them; but the most
splendid remains yet to be discovered, and the
victory is not hopeless; this is not the Blue Rose

but the climbing Cabbage Rose. A simple ama-
teur discovered the Rose du Roi; this ought to
give hope to every one who possesses a garden,
and a little leisure to cultivate the worship of
Flora.

A rural feast of some parts of France is called
the festival of the Roses, in which the best be-
haved maiden of the town or village is annually
crowned with Roses. The Persians have also an
annual festival of Roses, which consists of bands
of youths parading the streets with music, and
offering Roses to all they meet, for which they
receive a trifling gratuity. Rarities in Roses are
held in high estimation all over the world. At
the Botanical Garden of the East Indian Island
of Java, Dr. R. H. Scheffer, the Director, states
that there the Teas, Noisettes and Bourbons grow
well, and are always in bloom without ceasing.
The Hybrid Perpetuals flower best on the hills.
The rich Chinese residing in Java are great Rose
buyers, and do not mind paying 25 florins for a
young plant of the Green Rose, or for a Marshal
Neil.

War of the Roses in English history, the well known feuds that prevailed between the houses of York and Lancaster, are so called from the emblems adopted by their respective partisans; the adherents of the house of York having the white, those of Lancaster the red Rose, as their distinguishing symbol. These wars originated with the descendants of Edward the III, and after extending over a period of eighty years, during which England formed an almost uninterrupted scene of bloodshed and devastation, were finally put an end to by the victory of Henry Tudor, Earl of Richmond, over Richard III in 1485, the victor uniting in his own person the title of Lancaster through his mother, and that of York by his marriage with the daughter of Edward VI. Since that period the Rose has been the emblem of England,

> " Which once was doomed
> When civil discord braved the field.
> To grace the banner and the shield,"

as the thistle and shamrock are respectively the symbols of Scotland and Ireland.

MR. H. B. ELLWANGER'S MONOGRAPH OF ROSES OF AMERICAN ORIGIN, 1880.

Prairie Rose and Noisette Rose. These two classes have their origin in America, *R. rubifolia*, the *Prairie Rose*, seeds of which were sown by Samuel John Feast, of Baltimore, and the plants fertilized produced

Baltimore Belle, and *Queen of the Prairies;* their hardiness and vigorous growth make them of great value.

Anna Maria, raised by S. Feast, 1843; color blush or pale pink, full flowers, few thorns.

Anna Eliza (Williams), dark purple.

Baltimore Belle, S. J. Feast, 1843; white with blush centre, of good full form; has some Noisette blood, which makes it a little tender, but the most beautiful of the class.

Eva Corinne, pale blush.

Gem of the Prairies, raised by A. Burgess, New York, 1865; a hybrid between Queen of the Prairies and Madame Laffaye remontant; very crimson, blotched with white.

Jane, rosy-blush, double, and finely shaped.

King of the Prairies, S. Feast, 1843; pale rose.

Gracilis, W. Prince, 1845; varying in beauty.

Linnæan Hill Beauty, white or pale blush.

Madam Caradori Allen, S. Feast, 1843; bright pink, semi-double.

Milledgeville, pale blush, tinged with red.

Miss Grinnell, pale pink.

Mrs. Hovey, Joshua Pierce, Washington, pale blush flowers, becoming almost white, resembling Baltimore Belle.

Mrs. Pierce, J. Pruet, 1850; blush.

Pallida, S. Feast, 1843; blush resembling *superba*.

Perpetual Pink, S. Feast, 1843; rosy purple.

Pride of Washington, deep rose, small flowers.

Queen of the Prairies, S. Feast, 1843; pale rose, changing to blush.

Triomphant, J. Pierce, 1850; deep rose, double and compact.

Ranunculiflora, small blush flowers.

Superba, S. Feast, 1850; pale red, changing to blush.

These are the only ones now propagated.

Rosa Noisettiana, *Noisette Rose*, originated in Charleston, South Carolina; a group of vigorous growth, nearly hardy, and produces abundance of flowers.

America, C. C. Page, Washington, 1859; vigorous, flowers large, creamy yellow; a cross from Solfaterre or Saffrano.

Beauty of Greenmount, J. Pentland, Baltimore, 1854; rosy red.

Champney's Pink Cluster, Champney; vigorous, flowers pink, semi-double.

Cinderella, C. C. Page, 1859; rosy crimson.

Dr. Kane, Pentland, 1856; growth free, flowers large, golden yellow; a seedling from Cloth of Gold.

Isabella Gray, Andrew Gray, Charleston, 1854; flowers large, golden yellow, a seedling from Cloth of Gold.

Nasalina, A. Cook, 1872; flowers pink, flat form, very fragrant, a seedling from Desprez.

Tuseneltin, A. Cook, Baltimore, 1860; pale yellow, a seedling from Solfaterre.

Woodland Marguerite, J. Pentland, 1859; growth vigorous, flowers pure white.

BOURBON ROSES.

Charles Getz, A. Cook, 1871; a hybrid, vigorous, flowers pure white.

George Peabody, J. Pentland, 1857; growth moderate, color purplish crimson, a probable seedling from Paul Joseph.

Oplitz, A. Cook, 1871; a hybrid, color fiery red, a seedling from Gloiredes Rosamines.

Renno, A. Cook, 1868; named after Gen'l Renno of Philadelphia; deep pink.

Setina, P. Henderson, 1859; identical with Hermosa, but a sport of stronger growth.

BENGAL ROSE.

James Sprunt, Rev. J. M. Sprunt, 1856; like Cramoisie, superior, vigorous, excellent climber.

HYBRID PERPETUAL ROSES.

Belle Americaine, D. Bell, New York, deep pink, flowers small but of fine form.

Madame Boll, D. Boll, vigorous growth, flowers and foliage large, color Carmine rose, free bloomer, hardy.

Madam Trudeau, D. Boll, 1850; deep rose, double, and well formed.

Charles Cook, A. Cook, 1871; scarlet crimson.

Contina, A. Cook, 1871; rosy pink.

Defence, A. Cook, 1871; shiny red, camelia formed, spineless.

La Brillante, A. Cook, 1872; brilliant red, raised from Napoleon III.

Rosalina, A. Cook, 1871; rose color.

Souvenir of President Lincoln, A. Cook, 1869; dark velvety crimson.

TEA ROSES.

American Banner, G. Cartwright Dedham, Massachusetts, 1877; a sport from Bon Silene, flowers carmine, striped white.

Caroline Cook, A. Cook, 1871; colour pink, a seedling from Saffrano.

Cornelia Cook, A. Cook, 1855; flowers white, tinged with red, not a free bloomer, seedling from Devoniensis.

Desauter, A. Cook, 1855; colour flesh, seedling from Devoniensis.

Gen'l Washington, C. C. Page, 1860; rosy crimson.

Isabella Sprunt, Rev. J. M. Sprunt, 1855; sulphur yellow, a sport from Saffrano, very superior; the best.

Paradene, A. Cook, canary yellow, seedling from Pactole.

President, W. Paul, London, 1860; growth moderate, flowers large, rose color.

EXTRAORDINARY ROSES OF AMERICAN GROWTH GROWING AT THE AUGUSTA COTTON FACTORY, GEORGIA, MARCH, 1875.

A rose-tree, Cloth of Gold, trained to wall of a mill, growing in heavy black soil (loam); planted in 1847; stem 13 ins. diameter—three branches 5 to 6 ins. diameter; 60 feet high, 50 feet wide, and blooms abundantly every season; some unopened buds were sent to the writer by Mr. F. Cogin, superintendent of the mill in 1875, whose description of this wonderful Rose-tree has been confirmed to me by travellers from Georgia.

From California we have as follows: The famous Gold of Ophir Rose-tree on the Maddox farm in Eldorado Co., Cal., was recently destroyed by a violent wind. Its stem was 26 inches in circumference, and the shrub itself had grown around and over an oak fifty feet high, stopping in its upward growth only because there was nothing upon which it could climb higher. When in full bloom a splendid mass of golden flowers concealed the oak entirely from view with a blaze of glory, which many persons have travelled far to see. This rose was no doubt the Cromatella or Cloth of Gold, a hybrid of the Noisette Rose.

THE LANGUAGE OF THE ROSE.

A charming little book of a French lady, Madame Louise Leneveux, is entitled *Fleurs emblematiques*. We extract from the work of this lady (less her graceful style), the mysterious language that has been applied to the different species and varieties of the Rose.

A Rose—beauty.

Beauty the most brilliant, like the Rose, lasts but for a day.

The Wild Rose—simplicity.

Simplicity of heart and manners, not simplicity of intellect.

Hundred-leaved Rose, or *double Rose*—the Graces that accompany the Muses.

Without doubt, for the reason that when the painters and poets depict the Graces accompany-

ing Venus and Cupid, they represent them crowned with the myrtle; but when the Graces follow
the Muses, they crown them with roses.

Ever-blooming Rose—beauty of the freshest charms.

"A mes yeux vous serez toujours belle."

Not only are these Rose-trees covered with flowers all the season, but it is of all the kinds that
which exhales the sweetest perfume.

White Rose—silence.

Mystery is one of the charms of love. Harpocrates, the god of silence, is represented with a
rose in his hand and a finger on his lips.

Yellow Rose—inconstancy.

Man thinks he can unite unfaithfulness with
constancy. A French author in regard to unfaithfulness has said, "It is little when known;
unknown, nothing. The fool blubbers about it,
the prudent man says not a word."

The Cinnamon Rose—pompous splendor. All that
glitters is not gold.

In this rose allusion is made to its splendid
color, and its somewhat disagreeable odor.

Musk Rose—capricious beauty.

To be capricious indicates a feeble mind; this Rose being, as the author says, of very uncertain growth.

Fairy Rose—gentleness.

Gentleness is the grace of childhood. The Fairy, or Miss Lawrence's Rose, is small, pretty and graceful, which make it comparable to the gentleness with which nature has endowed infancy.

Moss Rose—voluptuous love. No more dreams of platonic love.

This charming variety of rose is well known, having been in cultivation for upwards of a century. It is the symbol of pleasure, being without thorns, and its prickles are not stinging.

Rose Bud—young girl.

Modesty should defend beauty, as the thorn protects the rose.

The Bouquet of Roses—according to Mdme. Charlotte de Latour, "gratitude is sweeter than the perfume of Roses, and power often more ephemeral than the beauty of flowers."

A Rose Bush on a grassy lawn—good company is profitable.

A Crown of Roses—recompense of Virtue; and is given to the deserving maidens at the Rose-feasts of the French villages.

A Rose leaf—I am never importunate. Although this signification accords with the history of the Academy of Amadan, and of Dr. Zeb, it agrees very little with that of Smindridi de Sybaris.

Thus finishes the nomenclature of the Rose, which furnishes some phrases to the mysterious language of flowers—a language useful to misfortune, to friendship, and to love.

Boitard.